Complex Numbers and Polar Curves for Pre-Calc and Trig

With Problems and Detailed Solutions

By Kathryn Paulk
Copyright © 2023

Updated 07/30/2024

Table of Contents

2

<u>Introduction</u>

This book introduces complex numbers and explains how they may be represented in rectangular and polar form.   How to switch from one form to the other is also explained.

Then, polar curves are introduced with a few simple examples. The general equations for limacons, spirals, roses, and straight lines are summarized which helps with graphing them. The last half of this book contains sample problems with detailed solutions.

<u>Imaginary Numbers (bi)</u>

Imaginary Numbers	
$i = \sqrt{-1}$ $i^2 = -1$	
$\sqrt{-3}$	$= \sqrt{(-1)\cdot(3)}$ $= \sqrt{-1}\cdot\sqrt{3}$ $= i\sqrt{3} \approx 1.73\,i$
$\sqrt{-32}$	$= \sqrt{(-1)\cdot(16)\cdot(2)}$ $= \sqrt{-1}\cdot\sqrt{16}\cdot\sqrt{2}$ $= 4i\sqrt{2} \approx 5.66\,i$

Imaginary Numbers – Ex. 1

i	$= i$
i^2	$= -1$
i^3	$= i^2 \cdot i = -i$
i^4	$= (i^2)^2 = (-1)^2 = 1$
i^5	$= i^4 \cdot i = (i^2)^2 \cdot i = (1)i = i$
i^{120}	$= (i^2)^{60} = (-1)^{60} = 1$
i^{121}	$= (i^2)^{60} \cdot i = (-1)^{60} \cdot i = i$
i^{122}	$= (i^2)^{61} = (-1)^{61} = -1$

Imaginary Numbers – Ex. 2	
$3i + 2i$	$= 5i$
$3i \cdot 2i$	$= 6i^2 = 6(-1) = -6$
$\dfrac{10i}{2i}$	$= \dfrac{10}{2} = 5$
$(2 + i)(5 + i)$	$= 10 + 2i + 5i + i^2$ $= 10 + 7i - 1 = 9 + 7i$
$\dfrac{2 + i}{3 + i}$	$= \dfrac{2 + i}{3 + i} \cdot \left(\dfrac{3 - i}{3 - i}\right)$ $= \dfrac{6 - 2i + 3i - i^2}{3^2 - i^2}$ $= \dfrac{6 + i + 1}{9 + 1} = \dfrac{7 + i}{10}$

Complex Numbers

Rectangular Form (a + bi)

Complex Numbers
Rectangular Format

Format of a complex number: $a + bi$

A complex number has two parts:

a = The real part

b = The imaginary part

Complex Numbers
Rectangular Coordinates
in the Argand Plane

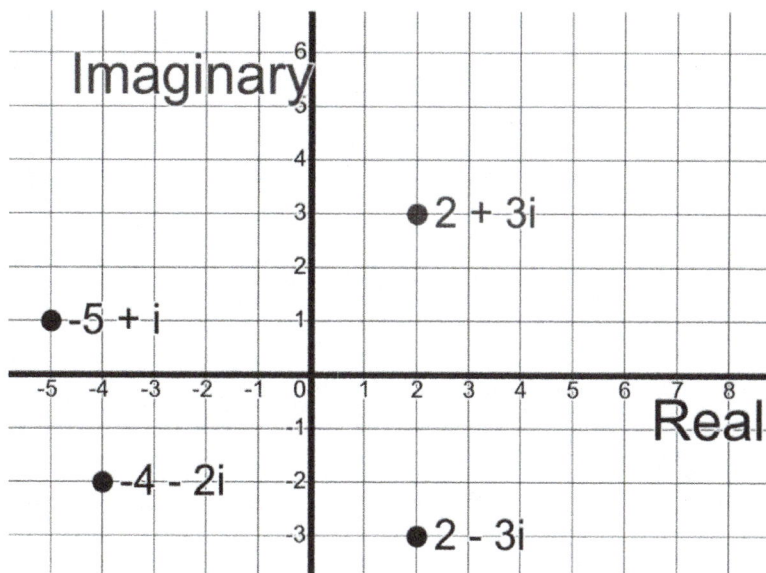

Imaginary

● 2 + 3i

● -5 + i

Real

● -4 - 2i

● 2 - 3i

Polar Form (cis)

Complex Numbers
Polar Form

Rectangular Form: $z = a + bi$

Polar Form: $\quad\quad z = r(\cos\theta + i\sin\theta)$

CIS Format: $\quad\quad z = r\,cis(\theta)$

Rectangular coordinates: $\quad (a, b)$

Polar coordinates: $\quad\quad\quad (r, \theta)$

Where:

$$a = r\cos\theta \quad\quad\quad b = r\sin\theta$$

$$r = |z| = \sqrt{a^2 + b^2} \quad\quad \text{(Modulus)}$$

$$\theta = \tan^{-1}\left(\frac{b}{a}\right) \quad\quad\quad \text{(Argument)}$$

Modulus = Absolute Value

For a complex number: $z = a + bi$

$|z|$ = Modulus (or Absolute Value)

$|z| = \sqrt{a^2 + b^2}$

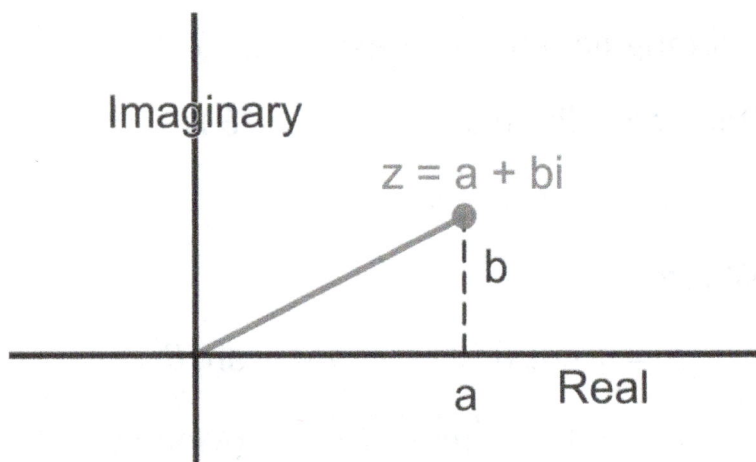

Polar Form – Ex. 1

Write the following numbers in polar form.

Number	Polar Form
$z = 1 + i$	$r = \lvert z \rvert = \sqrt{1^2 + 1^2} = \sqrt{2}$ $\theta = \tan^{-1}\left(\frac{1}{1}\right) = \frac{\pi}{4}$ $z = (r, \theta) = \left(\sqrt{2}, \frac{\pi}{4}\right)$ $z = \sqrt{2}\left(\cos\frac{\pi}{4} + i\sin\frac{\pi}{4}\right)$
$z = \sqrt{3} - i$	$r = \lvert z \rvert = \sqrt{\left(\sqrt{3}\right)^2 + 1^2} = 2$ $\theta = \tan^{-1}\left(\frac{-1}{\sqrt{3}}\right) = -\frac{\pi}{6}$ $z = (r, \theta) = \left(2, -\frac{\pi}{6}\right)$ $z = 2\left(\cos\left(-\frac{\pi}{6}\right) + i\sin\left(-\frac{\pi}{6}\right)\right)$

Polar Coordinate System

The Polar Coordinate System is a two-dimensional system used to graph points on a plane. Each point is identified by two polar coordinates (r, θ) where:

$r =$ distance from the origin

$\theta =$ the direction

$(r, \theta) = (5, 30^o)$	$(r, \theta) = (-5, 30^o)$
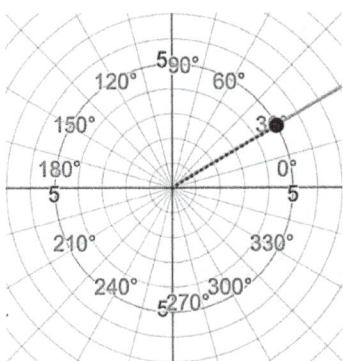	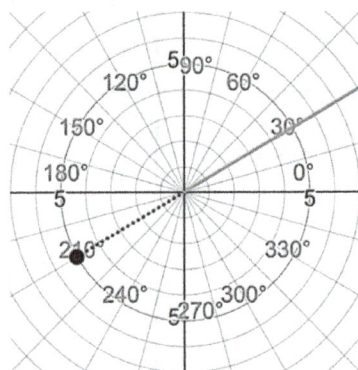

Polar Coordinate System
Examples With θ in Degrees

$(r,\theta) = (5,-30^o)$	$(r,\theta) = (-5,-30^o)$
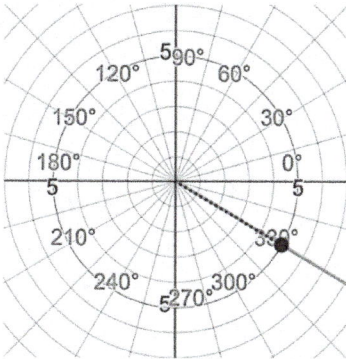	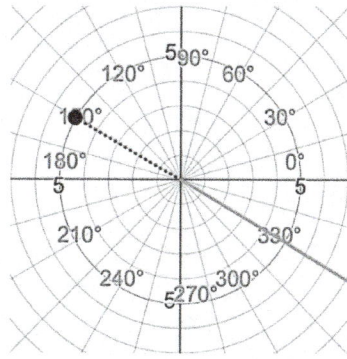

$(r,\theta) = (3,120^o)$	$(r,\theta) = (-3,120^o)$
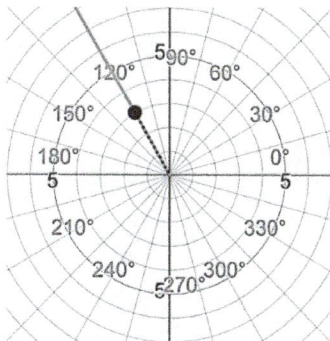	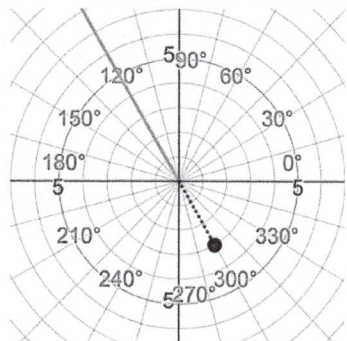

Polar Coordinate System
Examples With θ in Radians $\quad (\pi \approx 3.1)$

$(r, \theta) = (2, \pi)$	$(r, \theta) = (-2, \pi)$
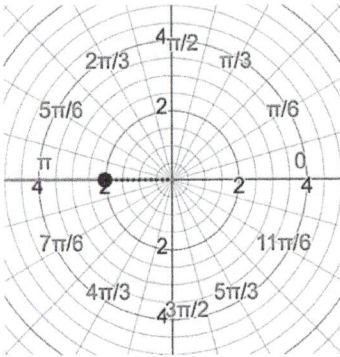	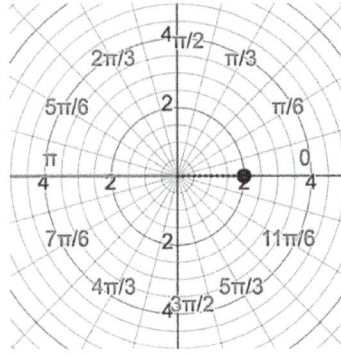
$(r, \theta) = (3, 2)$	$(r, \theta) = (-3, 2)$
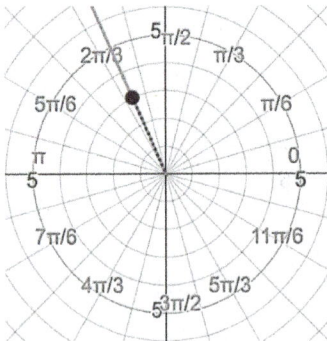	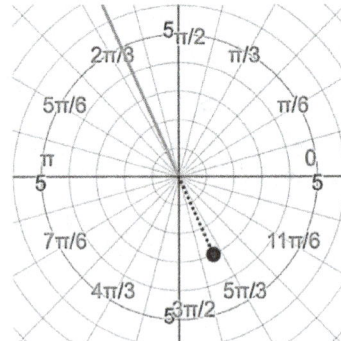

Conjugate Pairs

Properties of Conjugates

If: $z = a + bi$ and $w = c + di$

Then the conjugates are:

$\bar{z} = a - bi$ and $\bar{w} = c - di$

$\overline{z + w}$	$=$	$\bar{z} + \bar{w}$
$\overline{zw}$	$=$	$\bar{z} \cdot \bar{w}$
$\overline{z^n}$	$=$	$(\bar{z})^n$
$\lvert z \rvert$	$=$	$\sqrt{a^2 + b^2}$
$z \cdot \bar{z}$	$=$	$\lvert z \rvert^2$
$\dfrac{z}{w}$	$=$	$\dfrac{z}{w} \cdot \dfrac{\bar{w}}{\bar{w}} = \dfrac{z \cdot \bar{w}}{\lvert w \rvert^2}$

Conjugate Pairs – Ex. 1

Use the Quadratic Equation to find the roots of the equation: $x^2 + 2x + 2 = 0$

$$x = \frac{-b \pm \sqrt{b^2 - 4ac}}{2a}$$

$$x = \frac{-2 \pm \sqrt{2^2 - 4(1)(2)}}{2(1)}$$

$$x = \frac{-2 \pm \sqrt{4 - 8}}{2} = \frac{-2 \pm \sqrt{-4}}{2}$$

$$x = \frac{-2 \pm 2\sqrt{-1}}{2} = \frac{-2 \pm 2i}{2}$$

$$x = -1 \pm i \qquad \text{(Conjugate Pair)}$$

Conjugate Pairs – Ex. 2

Use the Quadratic Equation to find the roots of the equation: $x^2 + x + 1 = 0$

$$x = \frac{-b \pm \sqrt{b^2 - 4ac}}{2a}$$

$$x = \frac{-1 \pm \sqrt{1^2 - 4(1)(1)}}{2(1)}$$

$$x = \frac{-1 \pm \sqrt{1-4}}{2} = \frac{-1 \pm \sqrt{-3}}{2}$$

$$x = \frac{-1 \pm \sqrt{3}\, i}{2}$$

$$x = -\frac{1}{2} \pm \frac{\sqrt{3}}{2} i \qquad \text{(Conjugate Pair)}$$

Operations

Operations With Complex Numbers

Operations with complex numbers may be easier if the number is in Polar Form.

Complex Number	$z = a + bi$
Polar Form	$z = r\left(\cos\theta + i\sin\theta\right)$
Polar Form (CIS Format)	$z = r\,cis\,(\theta)$
Rectangular Coordinates	(a, b)
Polar Coordinates	(r, θ)

Where: $a = r\cos\theta$ $b = r\sin\theta$

$$r = \sqrt{a^2 + b^2} \qquad \text{(Modulus)}$$

$$\theta = \tan^{-1}\left(\frac{b}{a}\right) \qquad \text{(Argument)}$$

Addition and Subtraction

Adding Complex Numbers – Ex. 1

$(a + bi) + (c + di)$	$= (a + c) + (b + d)i$
$(2 + 3i) + (1 + 4i)$	$= (2 + 1) + (3 + 4)i$ $= 3 + 7i$
$(2 + 3i) + (1 - 4i)$	$= (2 + 1) + (3 - 4)i$ $= 3 - i$
$(2 + 3i) + (2 - 3i)$	$= (2 + 2) + (3 - 3)i$ $= 4$
$(0 - i) + (2 - 3i)$	$= (0 + 2) + (-1 - 3)i$ $= 2 - 4i$

Subtracting Complex Numbers – Ex. 1

$(a + bi) - (c + di)$	$= (a - c) + (b - d)\,i$
$(2 + 3i) - (1 + 4i)$	$= (2 - 1) + (3 - 4)\,i$ $= 1 - i$
$(2 + 3i) - (1 - 4i)$	$= (2 - 1) + (3 + 4)\,i$ $= 1 + 7i$
$(2 + 3i) - (2 - 3i)$	$= (2 - 2) + (3 + 3)\,i$ $= 6i$
$(0 - i) - (2 - 3i)$	$= (0 - 2) + (-1 + 3)\,i$ $= -2 + 2i$

Mult & Div (Rectangular Form)

Multiplying Complex Numbers
Rectangular Form – Ex. 1

$(a + bi) \cdot (c + di)$
$\quad = ac + adi + bci + bdi^2$

$(2 + 3i) \cdot (1 + 4i)$
$\quad = 2 + 8i + 3i + 12i^2$
$\quad = 2 + 11i + 12(-1)$
$\quad = 2 + 11i - 12 \ = -10 + 11i$

$(2 + 3i) \cdot (2 - 3i)$
$\quad = 4 - 6i + 6i - 9i^2$
$\quad = 4 - 9(-1) \ = \ 4 + 9 \ = \ 13$

$(0 - i) \cdot (2 - 3i)$
$\quad = -2i + 3i^2$
$\quad = -2i + 3(-1) = -3 - 2i$

Dividing Complex Numbers
Rectangular Form – Ex. 1

Use the difference of two squares
to simplify the denominator.

Recall: $a^2 - b^2 = (a - b)(a + b)$

$$(2 + 3i) \div (1 + 4i) \; = \; \frac{(2 + 3i)}{(1 + 4i)} \cdot \frac{(1 - 4i)}{(1 - 4i)}$$

$$= \frac{2 - 8i + 3i - 12i^2}{1^2 - (4i)^2}$$

$$= \frac{2 - 5i - 12(-1)}{1 - 16i^2}$$

$$= \frac{2 - 5i + 12}{1 - 16(-1)} = \frac{14 - 5i}{1 + 16} = \frac{14 - 5i}{17}$$

$$= \frac{14}{17} - \frac{5}{17}i$$

Mult & Div (Polar Form)

Multiplication and Division
Of Numbers in Polar Form

Given:

$$z_1 = (r_1, \theta_1) = r_1(\cos\theta_1 + i\sin\theta_1)$$
$$z_2 = (r_2, \theta_2) = r_2(\cos\theta_2 + i\sin\theta_2)$$

$z_1 \cdot z_2$	$= (r_1 \cdot r_2,\ \theta_1 + \theta_2)$
$\dfrac{z_1}{z_2}$	$= \left(\dfrac{r_1}{r_2},\ \theta_1 - \theta_2\right)$

Multiplication and Division
Polar Form – Ex. 1

Given: $z = (r, \theta) = \left(\sqrt{2}, \frac{\pi}{4}\right)$

$w = (r, \theta) = \left(2, -\frac{\pi}{6}\right)$

Find: The product $z \cdot w$ and quotient $\frac{z}{w}$

$z \cdot w$	$= (r_1 \cdot r_2,\ \theta_1 + \theta_2)$
	$= \left(2\sqrt{2},\ \frac{\pi}{4} + \left(-\frac{\pi}{6}\right)\right)$
	$= \left(2\sqrt{2},\ \frac{6\pi}{24} - \frac{4\pi}{24}\right) = \left(2\sqrt{2},\ \frac{\pi}{12}\right)$
$\frac{z}{w}$	$= \left(\frac{r_1}{r_2},\ \theta_1 - \theta_2\right)$
	$= \left(\frac{\sqrt{2}}{2},\ \frac{\pi}{4} - \left(-\frac{\pi}{6}\right)\right)$
	$= \left(\frac{\sqrt{2}}{2},\ \frac{\pi}{4} + \frac{\pi}{6}\right) = \left(\frac{\sqrt{2}}{2},\ \frac{5\pi}{12}\right)$

DeMoivre's Theorem (Exponents)

DeMoivre's Theorem

If $\quad z = (r, \theta) = r(\cos\theta + i\sin\theta)$

Then:

$$z^n = (r^n,\ n\theta)$$

$$z^n = r^n(\cos n\theta + i\sin n\theta)$$

$$n = \text{Positive Integer}$$

DeMoivre's Theorem – Ex. 1

Find: $\left(\dfrac{1}{2} + \dfrac{1}{2}i\right)^{10}$

Polar Form	$r = \sqrt{\left(\dfrac{1}{2}\right)^2 + \left(\dfrac{1}{2}\right)^2} = \sqrt{\dfrac{2}{4}} = \dfrac{\sqrt{2}}{2}$ $\theta = \tan^{-1}\left(\dfrac{1/2}{1/2}\right) = \dfrac{\pi}{4}$ $(r, \theta) = \left(\dfrac{\sqrt{2}}{2}, \dfrac{\pi}{4}\right)$

$(r, \theta)^n = (r^n, n\theta)$

$$\left(\dfrac{\sqrt{2}}{2}, \dfrac{\pi}{4}\right)^{10} = \left(\left(\dfrac{\sqrt{2}}{2}\right)^{10}, 10 \cdot \dfrac{\pi}{4}\right)$$

$$= \left(\dfrac{1}{32}, \dfrac{5\pi}{2}\right) = \dfrac{1}{32}i$$

Roots of Complex Numbers

Roots of a Complex Number

If $\quad z = (r, \theta) = r(\cos\theta + i\sin\theta)$

And $\quad n =$ Positive Integer

Then: $\quad z$ has n distinct nth roots.

$$(w_0, w_1, w_2, \ldots w_{n-1})$$

With: $\quad w_k = \left(r^{\frac{1}{n}}, \frac{\theta}{n} + \frac{2k\pi}{n} \right)$

$$k = 0, 1, 2, \ldots n - 1$$

n distinct roots

Roots of a Complex Number – Ex. 1a

Find the six 6th roots of $z = -8$

And graph the roots on the complex plane.

Polar Form	$z = -8 + 0i \ = \ \sqrt{8^2 + 0^2} \ = 8$
	$\theta = \tan^{-1}\left(\dfrac{0}{8}\right) = 0$
	$z = (r, \theta) = (8, \pi)$

$$w_k = \left(r^{\frac{1}{n}}, \ \frac{\theta}{n} + \frac{2k\pi}{n}\right) \qquad k = 0, 1, \ \ldots n - 1$$

$$w_0 = \left(8^{\frac{1}{6}}, \ \frac{\pi}{6} + \frac{2(0)\pi}{6}\right) = \left(\sqrt{2}, \ \frac{\pi}{6}\right)$$

$$w_1 = \left(8^{\frac{1}{6}}, \ \frac{\pi}{6} + \frac{2(1)\pi}{6}\right) = \left(\sqrt{2}, \ \frac{3\pi}{6}\right)$$

Continued ...

Roots of a Complex Number – Ex. 1b

$$w_k = \left(r^{\frac{1}{n}} , \; \frac{\theta}{n} + \frac{2k\pi}{n} \right) \qquad k = 0, 1, \; \ldots n - 1$$

$$w_0 = \left(8^{\frac{1}{6}} , \; \frac{\pi}{6} + \frac{2(0)\pi}{6} \right) = \left(\sqrt{2}, \; \frac{\pi}{6} \right)$$

$$w_1 = \left(8^{\frac{1}{6}} , \; \frac{\pi}{6} + \frac{2(1)\pi}{6} \right) = \left(\sqrt{2}, \; \frac{3\pi}{6} \right)$$

$$w_2 = \left(8^{\frac{1}{6}} , \; \frac{\pi}{6} + \frac{2(2)\pi}{6} \right) = \left(\sqrt{2}, \; \frac{5\pi}{6} \right)$$

$$w_3 = \left(8^{\frac{1}{6}} , \; \frac{\pi}{6} + \frac{2(3)\pi}{6} \right) = \left(\sqrt{2}, \; \frac{7\pi}{6} \right)$$

$$w_4 = \left(8^{\frac{1}{6}} , \; \frac{\pi}{6} + \frac{2(4)\pi}{6} \right) = \left(\sqrt{2}, \; \frac{9\pi}{6} \right)$$

$$w_5 = \left(8^{\frac{1}{6}} , \; \frac{\pi}{6} + \frac{2(5)\pi}{6} \right) = \left(\sqrt{2}, \; \frac{11\pi}{6} \right)$$

Continued …

Roots of a Complex Number – Ex. 1c

Find the six 6th roots of $z = -8$

And graph the roots on the complex plane.

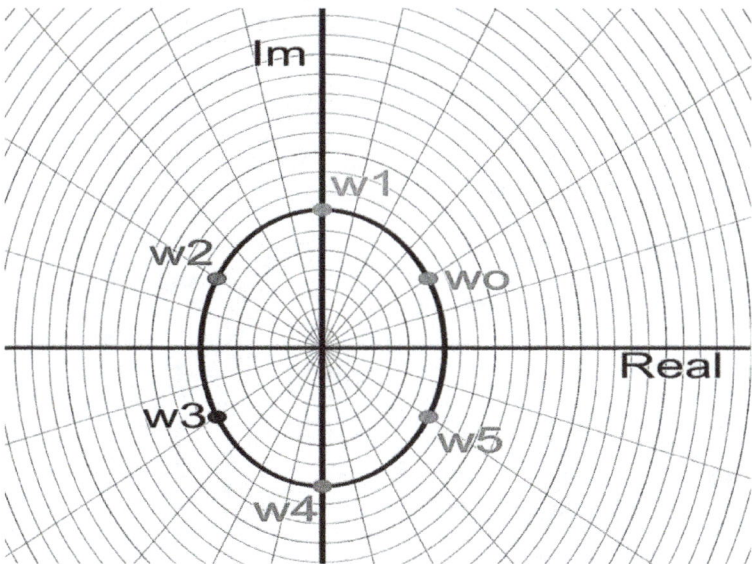

$$|z| = \sqrt{(-8)^2 + 0^2} = 8$$

Radius of circle $= 8^{\frac{1}{6}} = \sqrt{2} \approx 1.4$

Complex Exponents (Euler's Formula)

Complex Exponents   (Euler's Formula)

Euler's Formula: $e^{i\theta} = \cos\theta + i\sin\theta$

A complex number: $z = a + bi$
may be represented in different ways.

$$z = a + bi$$

$$z = (r, \theta)$$

$$z = r(\cos\theta + i\sin\theta)$$

$$z = r\ cis\ \theta$$

$$z = r \cdot e^{i\theta}$$

Where:

$$r = \sqrt{a^2 + b^2}$$

$$\theta = \tan^{-1}\left(\frac{b}{a}\right)$$

$$cis\ \theta = \cos\theta + i\sin\theta$$

Complex Exponents – Ex. 1

Evaluate: $e^{i\pi}$ and $e^{-1+\frac{i\pi}{2}}$

$e^{i\pi}$	$e^{i\pi} = \cos\pi + i\sin\pi$
	$= (-1) + i\,(0) \ = -1$

$e^{i\pi} + 1 = 0$ **Important Equation**

$e^{-1+\frac{i\pi}{2}}$

$$e^{-1+\frac{i\pi}{2}} = (e^{-1}) \cdot \left(e^{\frac{i\pi}{2}}\right)$$

$$= \left(\frac{1}{e}\right) \cdot \left(\cos\frac{\pi}{2} + i\,\sin\frac{\pi}{2}\right)$$

$$= \left(\frac{1}{e}\right) \cdot \left(0 + i\,(1)\right)$$

$$= \frac{i}{e}$$

Complex Exponents
Multiplication Example – Ex. 2a

Given: $z = 2 + 3i$ and $w = 1 + 4i$

Find: $z \cdot w$ Use complex exponents

z Polar Form	$r = \sqrt{2^2 + 3^2} = \sqrt{13}$ $\theta = \tan^{-1}\left(\frac{3}{2}\right) = 56.3^o$ $z = (r, \theta) = \left(\sqrt{13}, 56.3^o\right)$
w Polar Form	$r = \sqrt{1^2 + 4^2} = \sqrt{17}$ $\theta = \tan^{-1}\left(\frac{4}{1}\right) = 76^o$ $z = (r, \theta) = \left(\sqrt{17}, 76^o\right)$
$z \cdot w$	$z \cdot w = \sqrt{13}\, e^{i\,56.3} \cdot \sqrt{17}\, e^{i\,76}$ Continued ...

Complex Exponents
Multiplication Example – Ex. 2b

Given: $z = 2 + 3i$ and $w = 1 + 4i$

Find: $z \cdot w$ Use complex exponents

$z \cdot w$	
	$z \cdot w = \sqrt{13}\, e^{i\,56.3} \cdot \sqrt{17}\, e^{i\,76}$
	$= \sqrt{221}\, e^{i\,132.3}$
	$= \sqrt{221}\,(\cos 132.3 + i \sin 132.3)$
	$= \sqrt{221}\,\big((-.673) + i(.74)\big)$
	$= -10.0 + 11.0\,i$

Compare to multiplication, done previously.

$(2 + 3i) \cdot (1 + 4i)$

$\quad = 2 + 8i + 3i + 12i^2 = 2 + 11i + 12(-1)$

$\quad = 2 + 11i - 12 \quad = -10 + 11i$

Complex Exponents – Proof

$$e^{i\pi} + 1 \qquad = 0$$

$$cis(\pi) + 1 \qquad = 0$$

$$(\cos\pi + i\sin\pi) + 1 \qquad = 0$$

$$(-1 + 0) + 1 \qquad = 0$$

$$0 \qquad = 0$$

Polar Curves

Polar Curves

A polar equation: $r = f(\theta)$

Note: The radius may change, as θ changes.

Recall:

The complex number: $z = a + bi$

Has rectangular coordinates: (a, b)

And has polar coordinates: (r, θ)

For a fixed radius and angle.

Where:

$$a = r \cos \theta \qquad b = r \sin \theta$$

$$r = |z| = \sqrt{a^2 + b^2} \qquad \text{(Modulus)}$$

$$\theta = \tan^{-1}\left(\frac{b}{a}\right) \qquad \text{(Argument)}$$

$$z = (r, \theta) = r(\cos \theta + i \sin \theta)$$

Polar Curve Examples

Polar Curve – Ex. 1

Sketch the curve represented by $r = 2$

Make a T-table		

$r = 2$	θ
2	0
2	$\dfrac{\pi}{2}$
2	π
2	$\dfrac{3\pi}{2}$
2	2π

Plot the T-table points (r, θ)

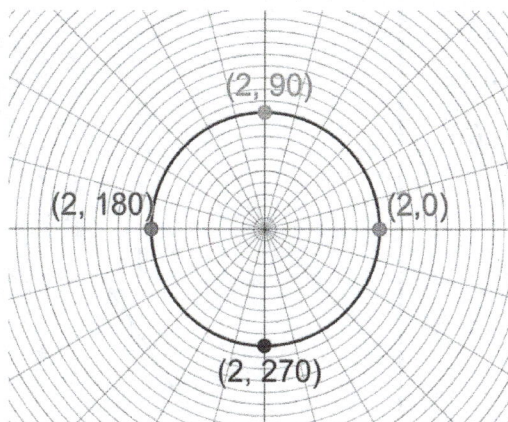

Polar Curve – Ex. 2

Sketch the curve represented by $\theta = 1$ rad

Note: The angle is constant

Make a T-table		
	$r = $ anything	θ
	0	1
	1	1
	2	1
	-1	1

Plot the T-table points (r, θ)

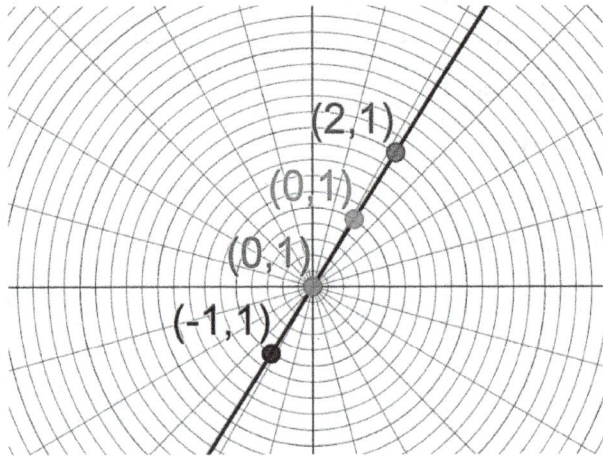

Polar Curve – Ex. 3

Sketch the curve represented by $r = 2\cos\theta$

Make a T-table	$r = 2\cos\theta$	θ
	2	0
	$\sqrt{2}$	45°
	0	90°
	-2	180°
	0	270°
	2	360°

Plot the T-table points (r, θ)

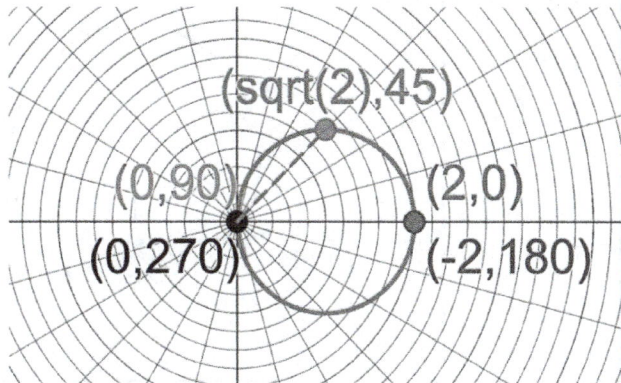

Polar Curve – Ex. 4

Sketch the curve: $r = 1 + \sin\theta$

Make a T-table		

$r = 1 + \sin\theta$	θ
1	0
2	90°
1	180°
0	270°
1	360°

Plot the
T-table
points
(r, θ)

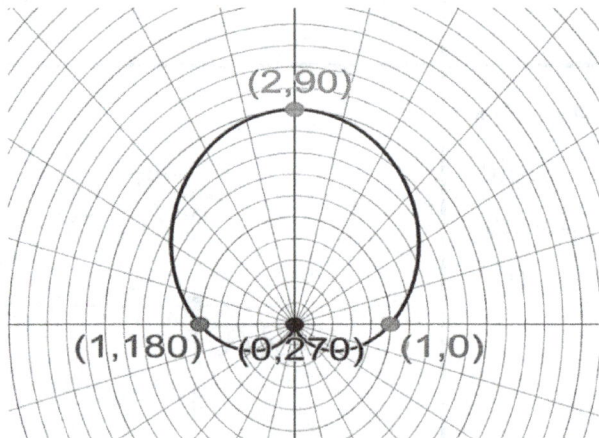

Polar Curve – Ex. 5

Sketch the curve: $r = \cos 2\theta$

Make a T-table	

$r = \cos 2\theta$	θ
1	0
0	45°
-1	90°
1	180°
-1	270°

Plot the T-table points (r, θ)

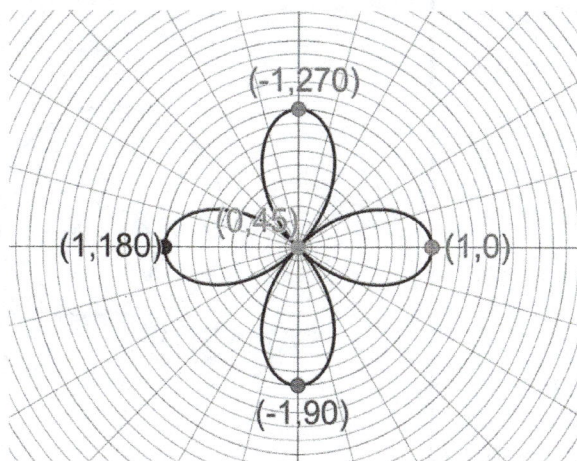

Polar Curve – Ex. 6

Sketch the curve: $r = 1 + 3\sin\theta$

Make a T-table		

$r = 1 + 3\sin\theta$	θ
1	0
4	90°
1	180°
-2	270°

Plot the T-table points (r, θ)

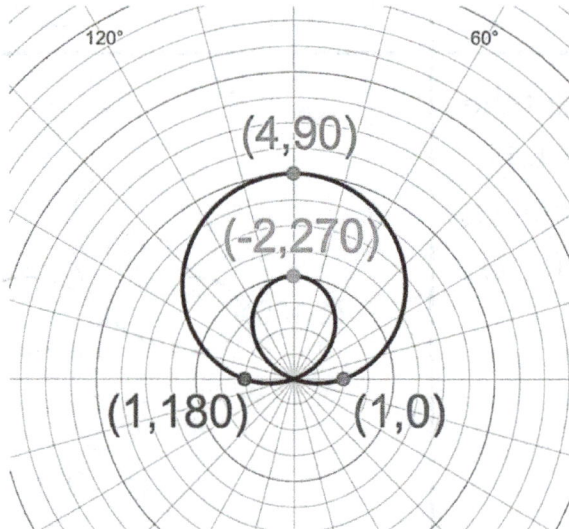

(4,90)

(-2,270)

(1,180) (1,0)

Polar Curve – Ex. 7

Sketch the curve: $r = 1 - 3\sin\theta$

Make a T-table	

$r = 1 - 3\sin\theta$	θ
1	0
-2	90°
1	180°
4	270°

Plot the T-table points (r,θ)

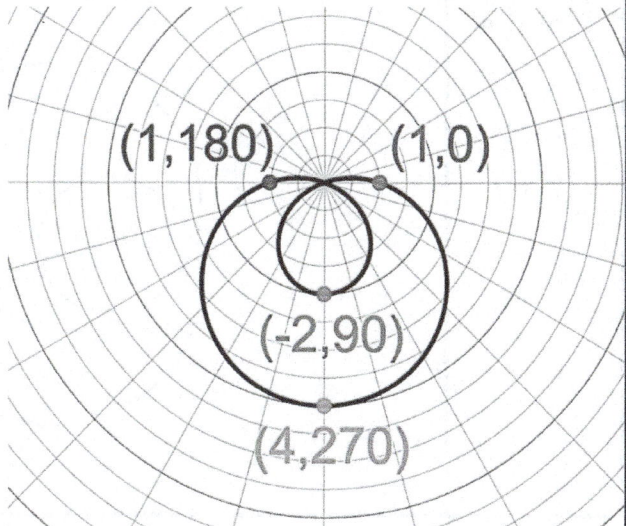

(1,180) (1,0)

(-2,90)

(4,270)

Polar Curve – Ex. 8

Sketch the curve: $r = 1 + 3\cos\theta$

Make a T-table		

$r = 1 + 3\cos\theta$	θ
4	0
1	90°
-2	180°
1	270°

Plot the T-table points (r, θ)

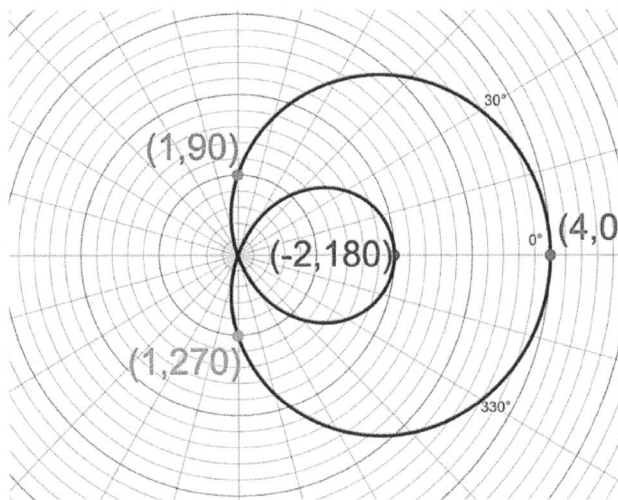

Polar Curve Patterns

Polar Curve Patterns

A good way to understand the patterns of polar curves is to use a graphing tool (like Desmos) and try different polar equations.

Desmos has a free APP and a free online graphing tool. When graphing polar curves with Desmos, click on the tool icon (a wrench) and select the polar grid. Also select radians or degrees.

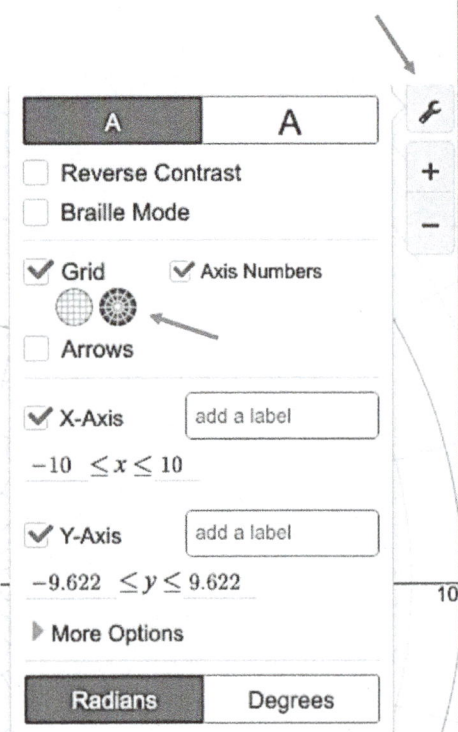

A	A

☐ Reverse Contrast
☐ Braille Mode

☑ Grid   ☑ Axis Numbers

☐ Arrows

☑ X-Axis   add a label
$-10 \leq x \leq 10$

☑ Y-Axis   add a label
$-9.622 \leq y \leq 9.622$

▶ More Options

Radians	Degrees

Polar Curve Patterns -- Overview

Curve	Equation	Example
Limacon	$r = a \pm b \sin\theta$ $r = a \pm b \cos\theta$	
Spiral	$r = a\theta$	
Circles	$r = a \sin\theta$ $r = a \cos\theta$	
Roses	$r = a \sin(n\theta)$ $r = a \cos(n\theta)$	

Polar Curve – Limacons
$$r = a + b\cos\theta$$

Here, the curve lines up along the x-axis.

$b = 0$	Just a constant radius for any θ. A perfect circle with radius = r.	
$a > b$	Perfect circle has a little dent in it. Often called a "bean."	
$a = b$	Perfect circle competes equally with the curving effect of the cos. Called a "heart" or "Cardioid."	
$a < b$	Perfect circle is overwhelmed with curving effect of cos.	

Polar Curve -- Limacons
$$r = a + b\sin\theta$$

Here, the curve lines up along the y-axis.

$b = 0$	Just a constant radius for any θ. A perfect circle with radius = r.	
$a > b$	Perfect circle has a little dent in it. Often called a "bean."	
$a = b$	Perfect circle competes equally with the curving effect of the sine. Called a "heart" or "Cardioid."	
$a < b$	Perfect circle is overwhelmed with curving effect of sine.	

Polar Curve -- Spiral
$r = n\theta$

$r = \theta$	Starts upward	
$r = -\theta$	Starts downward	
$r = 2\theta$	Starts upward	
$r = -2\theta$	Starts downward	

63

Polar Curve -- Roses
$r = a \, \text{trig}(n\theta)$ $n = $ Even

When n is even, $2n = $ the number of leaves

$r = 7\cos(2\theta)$	$r = 7\sin(2\theta)$
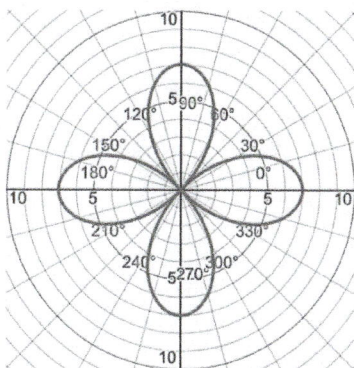	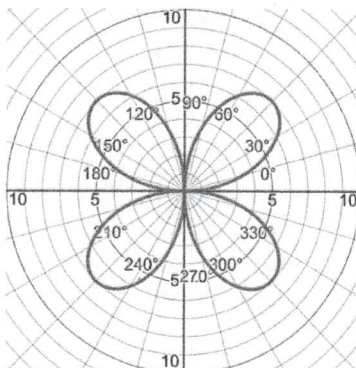

$r = 7\cos(4\theta)$	$r = 7\sin(4\theta)$
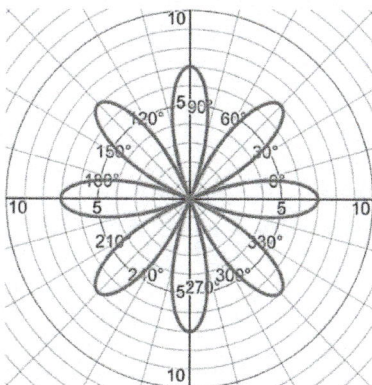	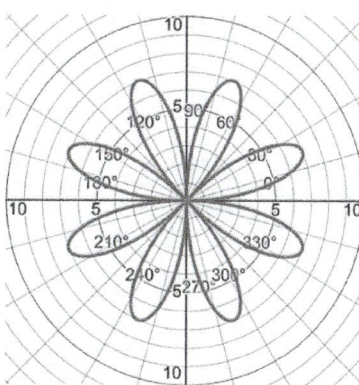

Polar Curve -- Roses
$r = a\,\mathrm{trig}(n\theta)$ $n = $ **Odd**

When n is odd, $n = $ the number of leaves

$r = 6\cos(3\theta)$

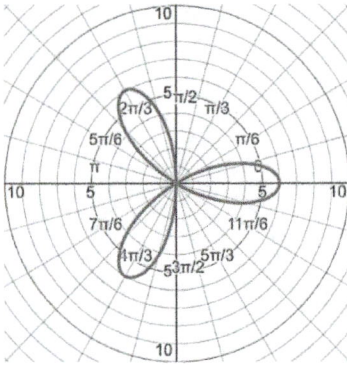

$r = 6\sin(3\theta)$

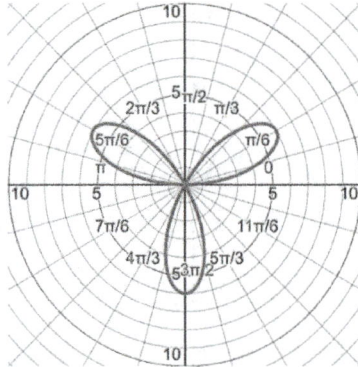

$r = 6\cos(5\theta)$

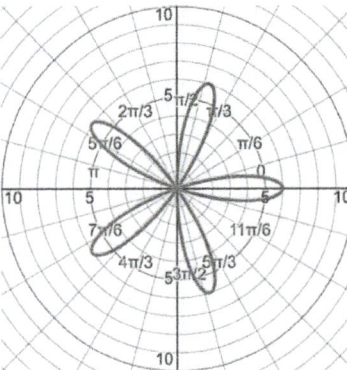

$r = 6\sin(5\theta)$

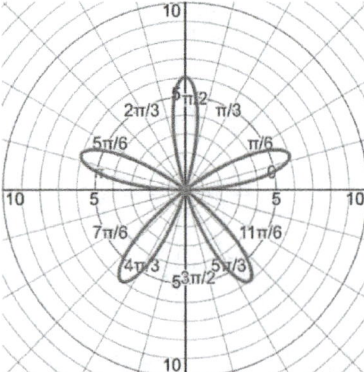

| Polar Curve -- Roses | | |
| $r = a\ \sin(n\theta)$ | | $n =$ **Odd** |
n	i^n	$r = \sin(n\theta)$
1	$i^1 = i$ Rose with 1 leaf	
3	$i^3 = (i^2)\,i$ $i^3 = (-1)\,i$	
5	$i^5 = (i^4)\,i$ $i^5 = (\,1\,)\,i$	
7	$i^7 = (i^6)\,i$ $i^7 = (-1)\,i$	
9	$i^9 = (i^8)\,i$ $i^9 = (\,1\,)\,i$	

Polar Curve – Straight Line
$\theta = constant$, r is not specified

When r is not specified, it can be anything.

$\theta = 30$

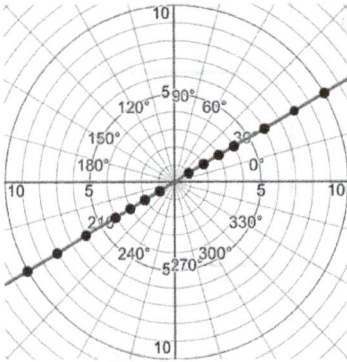

$\theta = -30$

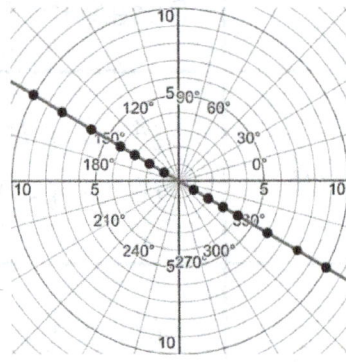

$\theta = 60$

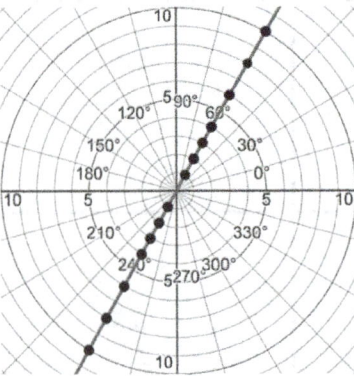

$\theta = -60$

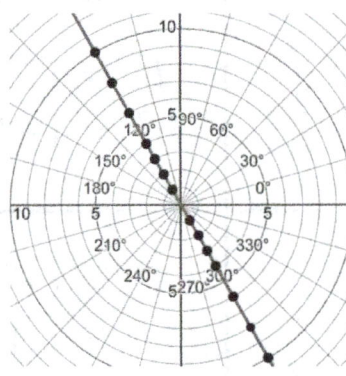

<u>Summary</u>

Summary – Complex Numbers	
Complex Numbers	$z = a + bi$ (Rectangular) $z = (r, \theta)$ (Polar) $z = r(\cos\theta + i\sin\theta)$ $z = r\ cis\ \theta$ $z = r \cdot e^{i\theta}$
Rect. To Polar	$r = \sqrt{a^2 + b^2}$ $\theta = \tan^{-1}\left(\dfrac{b}{a}\right)$

Summary – Polar Equations
$$r = f(\theta)$$

Circle	$r = 2$	
Spiral	$r = \left(\frac{1}{4}\right)\theta$	
Rose	$r = 2\sin(5\theta)$	
Limacon	$r = 2 + \sin\theta$ Bean	
	$r = 2 + 2\sin\theta$ Cardioid	
	$r = 1 + 2\sin\theta$	
	$r = 2\sin\theta$ Circle	

Summary – Limacon

$$r = a + b \sin \theta \qquad a < b$$

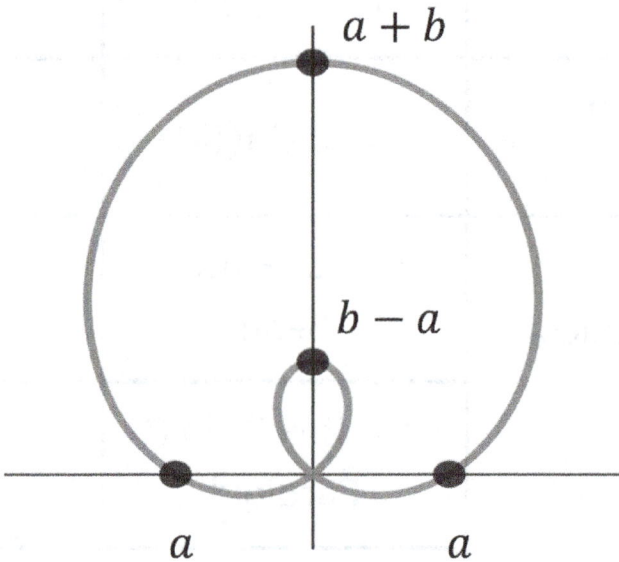

Sample Problems & Solutions

Convert Rectangular to Polar Equation

Convert Rectangular To Polar Form
Example 1

Given	$y = 2x - 5$
Substitute: $x = r \cos \theta$ $y = r \sin \theta$	$(r \sin \theta) = 2(r \cos \theta) - 5$
Solve for r	$r \sin \theta - 2r \cos \theta = -5$ $r(\sin \theta - 2 \cos \theta) = -5$ $r = \dfrac{-5}{(\sin \theta - 2 \cos \theta)}$
Note	$r = f(\theta)$ That's the goal!

Convert Rectangular To Polar Form Example 2	
Given	$y = \dfrac{\sqrt{3}}{3} x$
Note:	This is in the form $y = mx$ So, it's a straight line with slope $m = \dfrac{\sqrt{3}}{3}$
Use the slope to identify θ	$\theta = \tan^{-1}\left(\dfrac{\sqrt{3}}{3}\right) = \dfrac{\pi}{6}$
Polar Form of equation	$\theta = \dfrac{\pi}{6}$
Note:	Since r is not specified, it can be any value.

Convert Rectangular To Polar Form
Example 3

Given	$y^2 = 7x$
Substitute: $x = r\cos\theta$ $y = r\sin\theta$	$(r\sin\theta)^2 = 7(r\cos\theta)$ $r^2\sin^2\theta = 7r\cos\theta$
Solve for r	$r^2\sin^2\theta - 7r\cos\theta = 0$ $r(r\sin^2\theta - 7\cos\theta) = 0$ $r = 0$ (Trivial Case) OR $r = \dfrac{7}{\sin^2\theta}$
Note	$r = f(\theta)$ That's the goal!

Convert Rectangular To Polar Form Example 4	
Find polar coordinates for the rectangular point $(x, y) = (-3, -8)$ Where $r > 0$ and $360^o \leq \theta < 720^o$.	
r	$r = \sqrt{x^2 + y^2} = \sqrt{3^2 + 8^2} = \sqrt{73}$
θ	$\theta = \tan^{-1}\left(\frac{-8}{-3}\right) = 69.4^o$ θ must be in original quadrant Q3 $\theta = 69.4^o + 180^o = 249.4^o$ in Q3 θ must be in specified range. $\theta = 249.4^o + 360^o = 609.4^o$ in Q3
(r, θ)	$(r, \theta) = \left(\sqrt{73}, 609.4^o\right)$

Convert Rectangular To Polar Form
Example 5

Given	$x + y = 0$
Substitute: $x = r\cos\theta$ $y = r\sin\theta$	$r\cos\theta + r\sin\theta = 0$ $r(\cos\theta + \sin\theta) = 0$
Solve for r	$r = 0$ (Trivial Case) OR $\sin\theta = -\cos\theta$ $\tan\theta = -1 \quad \rightarrow \quad \theta = \dfrac{3\pi}{4}, \dfrac{7\pi}{4}$
Polar Eqn.	$\theta = \dfrac{3\pi}{4}$
Note:	Since r is not specified, it can be any value.

Convert Rectangular To Polar Form
Example 6

Given	$y = \frac{1}{3}x - 5$
Simplify	$3y = x - 15$
Substitute: $x = r\cos\theta$ $y = r\sin\theta$	$3r\sin\theta = r\cos\theta - 15$
Solve for r	$3r\sin\theta - r\cos\theta = -15$ $r(3\sin\theta - \cos\theta) = -15$ $r = \dfrac{-15}{3\sin\theta - \cos\theta}$
Polar Eqn.	$r = \dfrac{-15}{3\sin\theta - \cos\theta}$

Convert Polar to Rectangular Equation

Convert Polar to Rectangular Form
Example 1

Given	$r = -5\sec\theta$
Look for: $r\cos\theta = x$ $r\sin\theta = y$	$r = \dfrac{-5}{\cos\theta}$ $r\cos\theta = -5$ $x = -5$
Note	This is a vertical line at $x = -5$. Since y is not specified, it can be any value.

Convert Polar to Rectangular Form
Example 2

Given	$r = 3 \cos \theta$
Look for: $x = r \cos \theta$ $y = r \sin \theta$ $r^2 = x^2 + y^2$	$r \cdot (r) = r \cdot (3 \cos \theta)$ $r^2 = 3 (r \cos \theta)$ $x^2 + y^2 = 3x$ $x^2 + y^2 - 3x = 0$
Rectangular Form	$x^2 + y^2 - 3x = 0$

Convert Polar to Rectangular Form Example 3	
Given	$\theta = \dfrac{\pi}{4}$
Note:	Here, the angle is constant and $r =$ any value. It's a straight line with slope $= \tan\left(\dfrac{\pi}{4}\right) = 1$
Convert to x's and y's Use: $x = r\cos\theta$ $y = r\sin\theta$	$\tan(\theta) = 1$ $\dfrac{\sin\theta}{\cos\theta} = 1$ $\sin\theta = \cos\theta$ $r\sin\theta = r\cos\theta$ $y = x$ (Rectagular Form)

Convert Polar to Rectangular Form
Example 4

Given	$r = \dfrac{2}{1 - 3\sin\theta}$
Look for: $x = r\cos\theta$ $y = r\sin\theta$ r^2 $= x^2 + y^2$	$r(1 - 3\sin\theta) = 2$ $r - 3r\sin\theta = 2$ $r - 3y = 2$ $r = 2 + 3y$ $r^2 = (2 + 3y)^2$ $x^2 + y^2 = 4 + 12y + 9y^2$
Rectangular Form	$x^2 - 8y^2 - 12y - 4 = 0$

Convert Polar to Rectangular Form
Example 5

Given	$r = \dfrac{2}{3\cos\theta - 4\sin\theta}$
Look for: $x = r\cos\theta$ $y = r\sin\theta$	$r(3\cos\theta - 4\sin\theta) = 2$ $3r\cos\theta - 4r\sin\theta = 2$ $3x - 4y = 2$ $-4y = 2 - 3x$ $y = -\dfrac{2}{4} + \dfrac{3}{4}x$
Rectangular Form	$y = \dfrac{3}{4}x - \dfrac{1}{2}$

Convert Polar to Rectangular Form
Example 6

Polar (r, θ)	Rectangular (x, y)
$(5, 234^o)$	$x = 5\cos 234 \approx -2.9$ $y = 5\sin 234 \approx -4.0$ $(x, y) = (-2.9, -4.0)$
$\left(3, -\frac{11\pi}{6}\right)$	Use $\theta = -\frac{11\pi}{6} + 2\pi = \frac{\pi}{6}$ $x = 3\cos\left(\frac{\pi}{6}\right) = \frac{3\sqrt{3}}{2}$ $y = 3\sin\left(\frac{\pi}{6}\right) = \frac{3}{2}$ $(x, y) = \left(\frac{3\sqrt{3}}{2}, \frac{3}{2}\right)$

Graph the Given Polar Equation

Graph the Polar Equation
Example 1

Given	$r = -5$
Note:	The radius is always -5. It is the same for all θ. It is a circle with radius $= 5$
Polar Graph	

Graph the Polar Equation
Example 2

Given	$\theta = \dfrac{\pi}{3}$
Note:	The angle is always $\theta = \dfrac{\pi}{3}$. Radius is not specified so it can be any value. It is a straight line with Slope $= \tan\left(\dfrac{\pi}{3}\right) = \sqrt{3} \approx 1.7$
Polar Graph	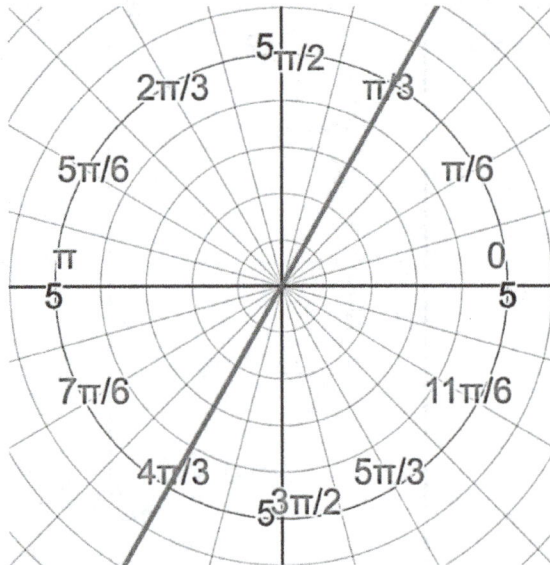

Graph the Polar Equation
Example 3

Given	$r = 5 + 4\cos\theta$
Note:	Lined up along x axis. Bean shape.

r	θ
9	0
5	90
1	180
5	270

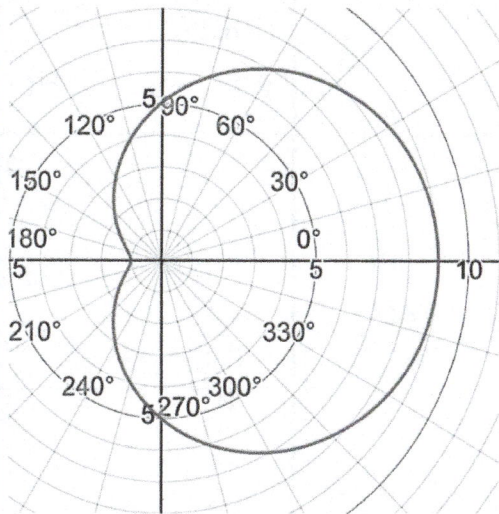

Graph the Polar Equation
Example 4

Given	$r = -5 \csc \theta$
Rearrange	$r = \dfrac{-5}{\sin \theta}$ $r \sin \theta = -5$ $y = -5 \qquad$ Horizontal Line
Polar Graph	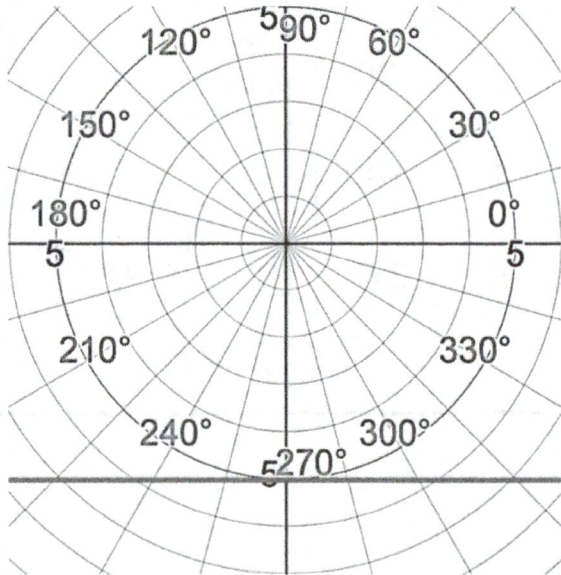

Write Polar Equation for Given Graph

Write Polar Equation for Given Graph
Example 1

Given Graph	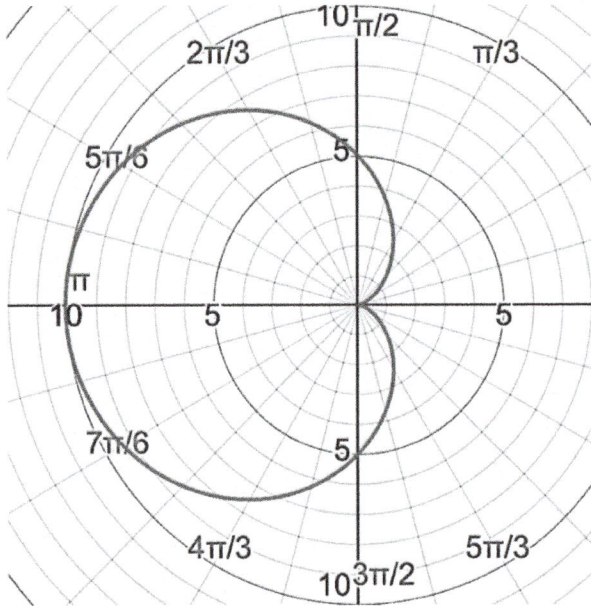
Note:	Lined up along x-axis on negative side so the form is: $r = a - b\cos\theta$ Heart shape, so $a = b$
Polar Eqn.	$r = 5 - 5\cos\theta$

Write Polar Equation for Given Graph
Example 2   (Radians) $\pi \approx 3.1$

Given Graph	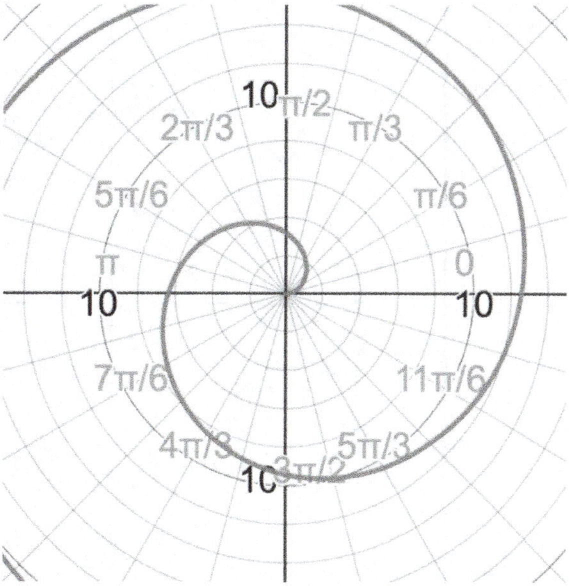
Note:	Spiral so in the form: $r = n\theta$ When $\theta = \pi$, $r \approx 6$ So $n = 2$ Starts in positive direction.
Polar Equation	$r = 2\theta$

Write Polar Equation for Given Graph
Example 3a (Radians) $\pi \approx 3.1$

Given Graph	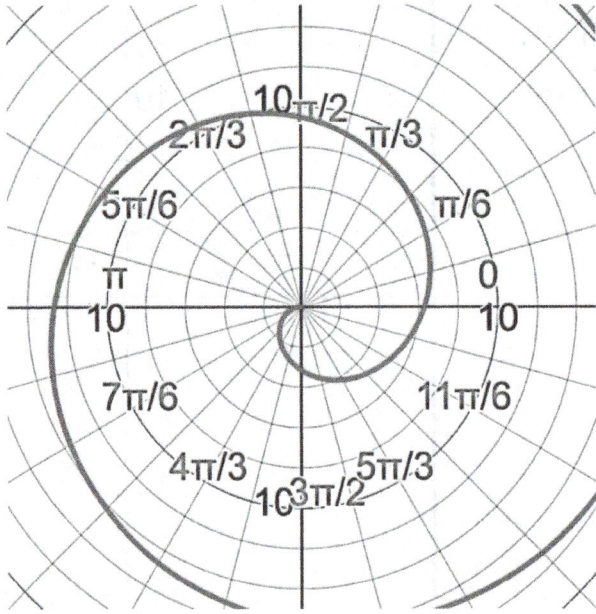
Note:	Spiral so in the form: $r = n\theta$ When $\theta = \pi$, $r \approx -6$ So $n = -2$ Starts in negative direction.
Polar Eqn.	$r = -2\theta$

Write Polar Equation for Given Graph
Example 3b   (Degrees)

Given Graph	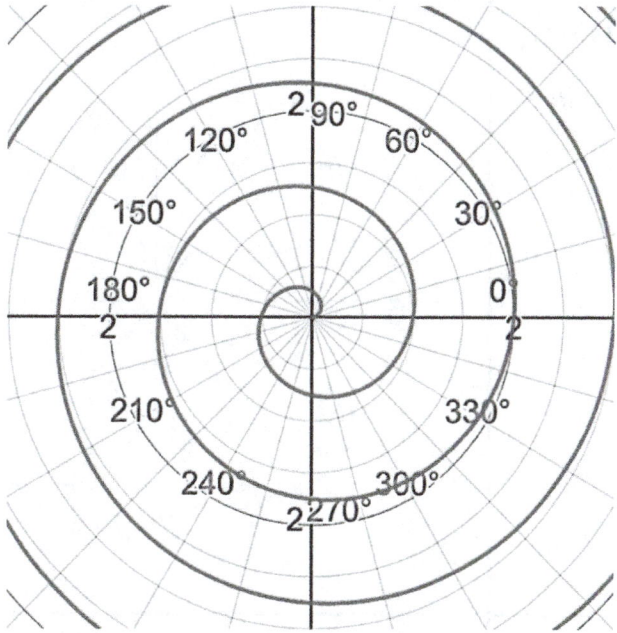
Note:	Spiral so in the form: $r = n\theta$ When $\theta = 360^o$, $r = 1$ So $n = \left(\dfrac{1}{360}\right)$ Starts in positive direction.
Polar Eqn.	$r = \left(\dfrac{1}{360}\right)\theta$

Write Polar Equation for Given Graph
Example 4

Given Graph	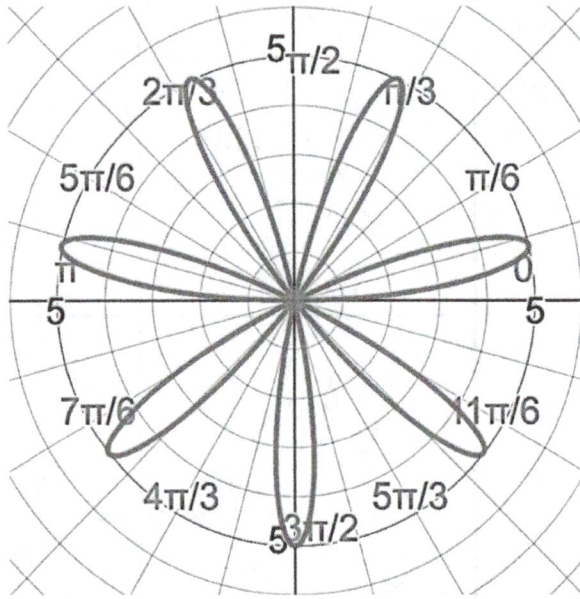
Note:	$r = a\sin(n\theta)$ with n = Odd So there are $n = 7$ pedals. Lined up with y-axis. Has y-axis symmetry.
Polar Eqn.	$r = 5\sin(7\theta)$

Write Polar Equation for Given Graph
Example 5

Given Graph	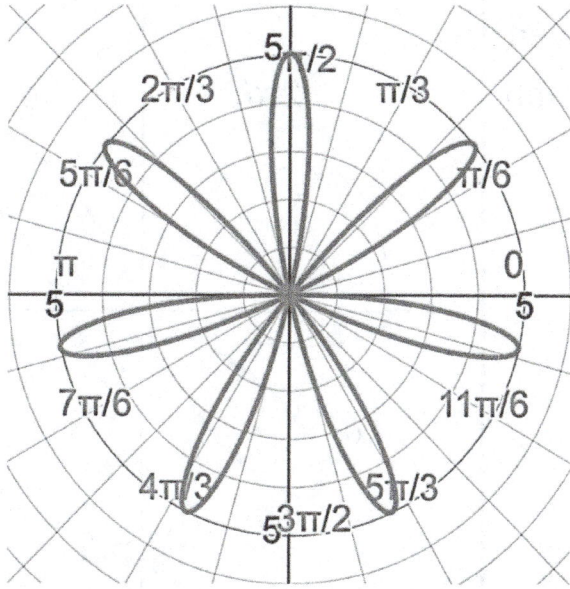
Note:	$r = a\sin(n\theta)$ with n = Odd So there are $n = 7$ pedals. Lined up with y-axis. Has y-axis symmetry.
Polar Eqn.	$r = -5\sin(7\theta)$

Write Polar Equation for Given Graph
Example 6

Given Graph	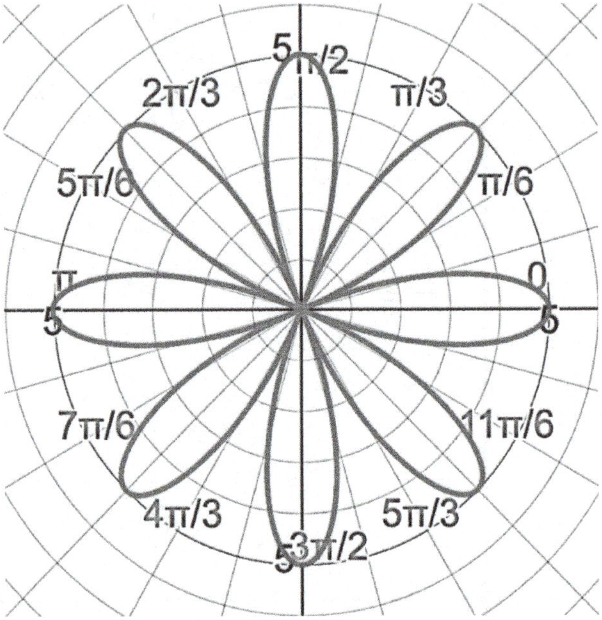
Note:	$r = a\cos(n\theta)$ with n = Even So there are $2n = 8$ pedals. $n = 4$ Lined up with x-axis. (and y-axis)
Polar Eqn.	$r = 5\cos(4\theta)$ or $r = -5\cos(4\theta)$ It has x-axis and y-axis symmetry.

Write Polar Equation for Given Graph
Example 7

Given Graph	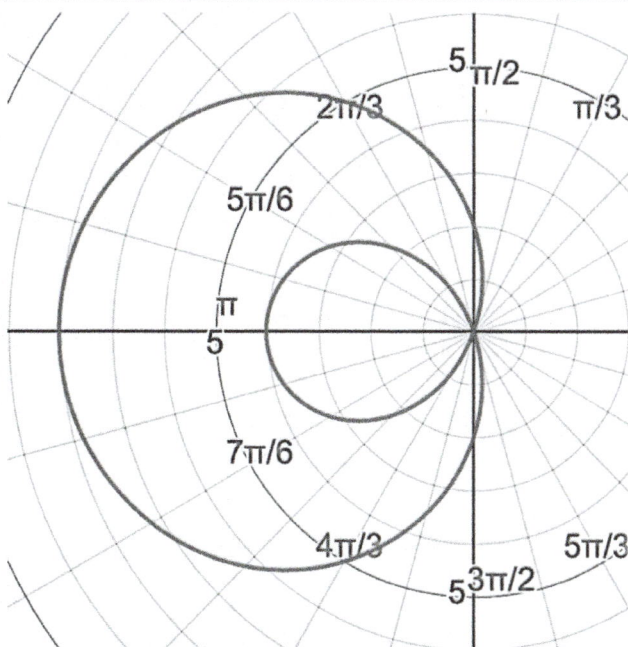
Note:	$r = -a - b\cos(\theta)$ Format With $|a| < |b|$
Polar Eqn.	$r = -2 - 6\cos(\theta)$

Find Domain, Range and Period

Find Domain, Range and Period Example 1	
Find the domain, range, and period of $$y = 5\sin(2x) + 3$$	
Domain	No restrictions on x. D: $(-\infty, \infty)$
Range	Min $= 5(-1) + 3 = -2$ Max $= 5(1) + 3 = 8$ R: $[min,\ max] = [-2,\ 8]$
Period	Horizontal compression. Period $= \frac{1}{2}(2\pi) = \pi$

Find Simultaneous Solutions

Find Simultaneous Solution(s)
Given 2 Polar Curves – Ex. 1a

Given Graphs	$r_1 = 4\sin 2\theta$ $r_2 = 4\cos\theta$
Make a Sketch	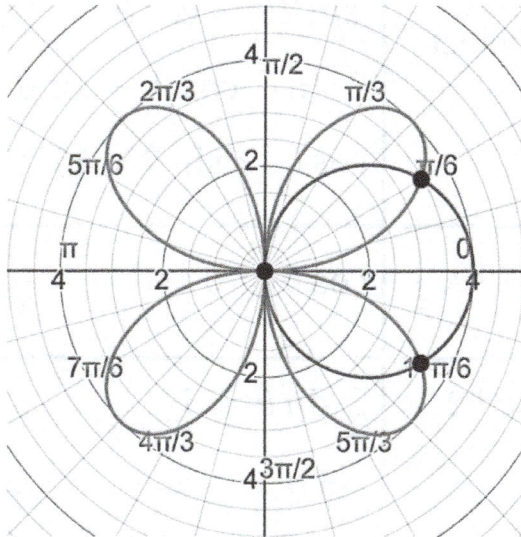
Notes	Looks like there are three points where the two curves are equal. Now, let's do the math!
Continued ...	

Find Simultaneous Solution(s)
Given 2 Polar Curves – Ex. 1b

Find where graphs intersect.	$r_1 = r_2$ $4 \sin 2\theta = 4 \cos \theta$ $\sin 2\theta = \cos \theta$ $2 \sin \theta \, \cos \theta - \cos \theta = 0$ $\cos \theta \, (2 \sin \theta - 1) = 0$
Case #1	$\cos \theta = 0 \;\rightarrow\; \theta = \dfrac{\pi}{2}, \dfrac{3\pi}{2}$
Case #2	$2 \sin \theta = 1$ $\sin \theta = \dfrac{1}{2}$ $\theta = \dfrac{\pi}{6}, \dfrac{5\pi}{6}$
Continued ...	

Find Simultaneous Solution(s)
Given 2 Polar Curves – Ex. 1c

Note: $r = r_1 = r_2 = 4\cos\theta$

θ	(r,θ)	$x = r\cos\theta$	$y = r\sin\theta$
$\dfrac{\pi}{2}$	$\left(0, \dfrac{\pi}{2}\right)$	0	0
$\dfrac{3\pi}{2}$	$\left(0, \dfrac{3\pi}{2}\right)$	0	0
$\dfrac{\pi}{6}$	$\left(2\sqrt{3}, \dfrac{\pi}{6}\right)$	3	$\sqrt{3} \approx 1.7$
$\dfrac{5\pi}{6}$	$\left(-2\sqrt{3}, \dfrac{5\pi}{6}\right)$	3	$-\sqrt{3}$

The 3 intersection points are:

$$(x,y) = (0,0), \left(3, \sqrt{3}\right), \left(3, -\sqrt{3}\right)$$

Convert Format of Complex Number

Convert Format of Complex Number Example 1

Given the Complex Number: $5 - 7i$

Convert it to Polar Complex in degree.

Note	Original point is in Q4
Find r	$r = \sqrt{a^2 + b^2}$ $r = \sqrt{5^2 + 7^2} = \sqrt{74}$
Find θ	$\theta = \tan^{-1}\left(\frac{-7}{5}\right) = -54.5$ (Q4) Make it a positive angle. $\theta = -54.5 + 360 = 305.5^o$
Polar Complex	$z = \sqrt{74}\ cis\ (305.5^o)$ $z = \sqrt{74}\ (\cos(305) + i\sin(305)$

Convert Format of Complex Number Example 2	
Convert $6\left(\cos\frac{5\pi}{6} + i\sin\frac{5\pi}{6}\right)$ to Rectangular Form.	
Find x	$x = r\cos\theta$ $x = 6\cos\frac{5\pi}{6} = 6\left(-\frac{\sqrt{3}}{2}\right)$ $x = -3\sqrt{3}$
Find y	$y = r\sin\theta$ $y = 6\sin\frac{5\pi}{6} = 6\left(\frac{1}{2}\right)$ $y = 3$
Rect. Form	$(x, y) = (-3\sqrt{3}, \ 3)$

Convert Format of Complex Number Example 3	
Convert $5 - 5i\sqrt{3}$ to Polar Complex	
Note	Original point is in Q4
Find r	$r = \sqrt{a^2 + b^2} = \sqrt{5^2 + \left(5\sqrt{3}\right)^2}$ $r = \sqrt{25 + 25 \cdot 3} = \sqrt{100} = 10$
Find θ	$\theta = \tan^{-1}\left(\dfrac{b}{a}\right)$ $\theta = \tan^{-1}\left(\dfrac{5\sqrt{3}}{5}\right) = \dfrac{\pi}{3}$ Q1 $\theta = -\dfrac{\pi}{3} + 2\pi = \dfrac{5\pi}{3}$ Q4
Polar Cpx.	$z = 10\, cis\left(\dfrac{5\pi}{3}\right)$ $z = 10\left(\cos\dfrac{5\pi}{3} + i\sin\dfrac{5\pi}{3}\right)$

<u>Mult, Divide & Add Complex Nums</u>

Multiply and Divide
Complex Numbers -- Example 1

Given:	$m = 3(\cos 56^o + i \sin 56^o)$
	$n = 4(\cos 78^o + i \sin 78^o)$
Find:	$m \cdot n$ and $\dfrac{m}{n}$

$m \cdot n$	$= 3\, cis(56) \cdot 4\, cis(78)$ $= (3 \cdot 4)\, cis(56 + 78)$ $= 12\, cis(134)$ $= 12(\cos 134 + i \sin 134)$
$\dfrac{m}{n}$	$= 3\, cis(56) \div 4\, cis(78)$ $= \left(\dfrac{3}{4}\right) cis\,(56 - 78)$ $= .75\, cis\,(-22 + 360)$ $= .75\, cis\,(338)$ $= .75\,(\cos 338 + i \sin 338)$

Subtract
Complex Numbers -- Example 2

Given:	$m = 3(\cos 56^o + i \sin 56^o)$
	$n = 4(\cos 78^o + i \sin 78^o)$
Find:	$m - n$

Real	$3\cos 56 - 4\cos 78 = 5.99$	
Imag.	$3\sin 56 - 4\sin 78 = -3.62$	
$m - n$	$= a + bi = 5.99 - 3.62\,i$	Q4
r	$= \sqrt{(5.99)^2 + (3.62)^2} = 48.98$	
θ	$= \tan^{-1}\left(\frac{-3.62}{5.99}\right) = -31.15$	Q4
	$= -31.15 + 360 = 328.9^o$	Q4
$m - n$	$= 48.98\; cis\,(328.9^o)$	

<u>Find Exponents of Complex Numbers</u>

Exponent of Complex Number -- Example 1
Given: $m = 3(\cos 56^o + i \sin 56^o)$ Find: m^6

m^6	$= [3\ cis\ (56)]^6$
	$= 3^6\ cis\ (6 \cdot 56)$
	$= 729\ cis\ (336)$
	$= 729\ (\cos 336 + i \sin 336)$

Exponent of
Complex Number -- Example 2

Given: $m = 3(\cos 56^o + i \sin 56^o)$

Find: m^6 Use Exponential Form

m^6	
	$= [3 \; cis \; (56)]^6$
	$= \left[3 \; e^{56 \cdot i}\right]^6$
	$= 3^6 \; e^{6 \cdot 56i}$
	$= 729 \; e^{336 \cdot i}$
	$= 729 \; cis \; (336)$
	$= 729 \, (\cos 336 + i \sin 336)$

Exponent of
Complex Number -- Example 3

| Given: | $z = \sqrt{2} - \sqrt{2}\,i$ | Q4 |
| Find: | z^5 | |

r	$= \sqrt{(\sqrt{2})^2 + (\sqrt{2})^2} = \sqrt{4} = 2$
θ	$= \tan^{-1}\left(\frac{-\sqrt{2}}{\sqrt{2}}\right) = \tan^{-1}(-1)$ $= -\frac{\pi}{4} = 2\pi - \frac{\pi}{4} = \frac{7\pi}{4}$ $\quad$ Q4
z^5	$= \left[2\,cis\left(\frac{7\pi}{4}\right)\right]^5$ $= 2^5\,cis\left(5 \cdot \frac{7\pi}{4}\right) = 32\,cis\left(\frac{35\pi}{4}\right)$ $= 32\,cis\left(\frac{32\pi}{4} + \frac{3\pi}{4}\right) = 32\,cis\left(\frac{3\pi}{4}\right)$ $= 32\left(\cos\frac{3\pi}{4} + i\sin\frac{3\pi}{4}\right)$

Find Roots of Complex Numbers

Roots of
Complex Number -- Example 1

Given: $\quad z = 16(\cos 100^o + i \sin 100^o)$

Find the 4th roots of z in polar form.

$z^{\frac{1}{4}}$	$= [16\ cis(100)]^{\frac{1}{4}}$
	$= 16^{\frac{1}{4}}\ cis\left(\frac{100}{4} + k \cdot \frac{360}{4}\right)$
	$= 2\ cis(25 + 90 \cdot k)\ ; \quad k = 0, 1, 2, 3$

$z_0 = 2\ cis(25 + 90 \cdot 0) = 2\ cis(25^o)$

$z_1 = 2\ cis(25 + 90 \cdot 1) = 2\ cis(115^o)$

$z_2 = 2\ cis(25 + 90 \cdot 2) = 2\ cis(205^o)$

$z_3 = 2\ cis(25 + 90 \cdot 3) = 2\ cis(295^o)$

Diagram (Extra)	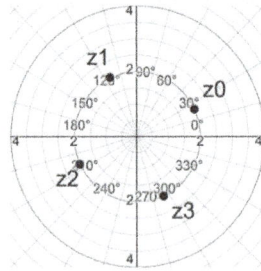

Roots of
Complex Number -- Example 2a

Given: $z = -27i$

Find the 3rd roots of in polar form. Radians.

r	$r = \sqrt{a^2 + b^2} = \sqrt{0^2 + 27^2} = 27$
θ	$\theta = \tan^{-1}\left(\frac{-27}{0}\right) = Undefined$ $\cos\theta = 0 \rightarrow \theta = \frac{\pi}{2}, \frac{3\pi}{2}$ $\quad$ Use $\frac{3\pi}{2}$
z	$z = r\,cis(\theta) = 27\,cis\left(\frac{3\pi}{2}\right)$
$z^{\frac{1}{3}}$	$= \left[27\,cis\left(\frac{3\pi}{2}\right)\right]^{\frac{1}{3}}$ $= 27^{\frac{1}{3}}\,cis\left(\frac{1}{3}\cdot\frac{3\pi}{2} + k\cdot\frac{2\pi}{3}\right)$ $= 3\,cis\left(\frac{\pi}{2} + \frac{2\pi}{3}\cdot k\right) ; \quad k = 0, 1, 2$

Continued ...

Roots of
Complex Number -- Example 2b

Given: $z = -27i$

Find the 3rd roots of in polar form. Radians.

z	$z = r\,cis(\theta) \;=\; 27\,cis\left(\frac{3\pi}{2}\right)$
$z^{\frac{1}{3}}$	$= \left[27\,cis\left(\frac{3\pi}{2}\right)\right]^{\frac{1}{3}}$
	$= 27^{\frac{1}{3}}\,cis\left(\frac{1}{3}\cdot\frac{3\pi}{2} \;+\; k\cdot\frac{2\pi}{3}\right)$
	$= 3\,cis\left(\frac{\pi}{2} \;+\; \frac{2\pi}{3}\cdot k\right) ; \quad k = 0,1,2$

$$z_0 = 3\,cis\left(\frac{3\pi}{6} + \frac{4\pi}{6}\cdot 0\right) = 3\,cis\left(\frac{3\pi}{6}\right)$$

$$z_1 = 3\,cis\left(\frac{3\pi}{6} + \frac{4\pi}{6}\cdot 1\right) = 3\,cis\left(\frac{7\pi}{6}\right)$$

$$z_2 = 3\,cis\left(\frac{3\pi}{6} + \frac{4\pi}{6}\cdot 2\right) = 3\,cis\left(\frac{11\pi}{6}\right)$$

Continued ...

Roots of
Complex Number -- Example 2c

Given: $z = -27i$

Find the 3rd roots of in polar form. Radians.

z	$z = r\,cis(\theta) = 27\,cis\left(\dfrac{3\pi}{2}\right)$
$z^{\frac{1}{3}}$	$= 3\,cis\left(\dfrac{1}{3}\cdot\dfrac{3\pi}{2} + \dfrac{2\pi}{3}\cdot k\right) ; \quad k = 0,1,2$

$$z_0 = 3\,cis\left(\frac{3\pi}{6} + \frac{4\pi}{6}\cdot 0\right) = 3\,cis\left(\frac{3\pi}{6}\right)$$

$$z_1 = 3\,cis\left(\frac{3\pi}{6} + \frac{4\pi}{6}\cdot 1\right) = 3\,cis\left(\frac{7\pi}{6}\right)$$

$$z_2 = 3\,cis\left(\frac{3\pi}{6} + \frac{4\pi}{6}\cdot 2\right) = 3\,cis\left(\frac{11\pi}{6}\right)$$

Diagram (Extra)	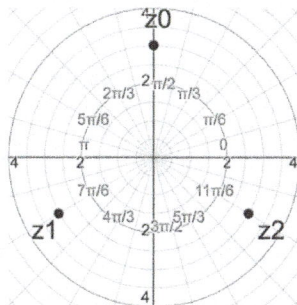

Locate Inner Loop of Cardioid

Locate Inner Loop of Cardioid
Example 1a

Given: $r = -\sqrt{3} + 2\sin\theta$

Find where the inner loop is formed. Radians.

Note	$r = a + b\sin\theta$ With $a < b$
	Loop occurs between where $r = 0$
Set $r = 0$	$-\sqrt{3} + 2\sin\theta = 0$
	$\sin\theta = \dfrac{\sqrt{3}}{2} \;\rightarrow\; \theta = \dfrac{\pi}{3}, \dfrac{2\pi}{3}$
Answer	Inner Loop occurs when:
	$\dfrac{\pi}{3} \le \theta \le \dfrac{2\pi}{3}$

See graph next page.

124

Locate Inner Loop of Cardioid
Example 1b

Given: $r = -\sqrt{3} + 2\sin\theta$

Find where the inner loop is formed. Radians.

Answer	Inner Loop occurs when: $\dfrac{\pi}{3} \le \theta \le \dfrac{2\pi}{3}$
Graph (Extra)	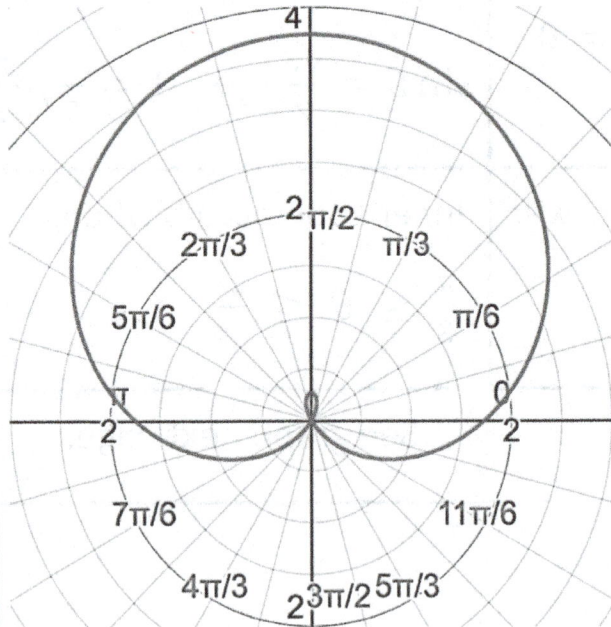

Locate Inner Loop of Cardioid
Example 1c

Given: $r = -\sqrt{3} + 2\sin\theta$

Find where the inner loop is formed. Radians.

Answer	Inner Loop occurs when:
	$$\frac{\pi}{3} \le \theta \le \frac{2\pi}{3}$$

Table (Extra)	$r = -\sqrt{3} + 2\sin\theta$	θ
	-1.7	0
	0	$\frac{\pi}{3}$
	1.6	$\frac{\pi}{2}$
	0	$\frac{2\pi}{3}$
	-1.7	π
	-3.5	$\frac{4\pi}{3}$
	-3.7	$\frac{3\pi}{2}$

Other Books
by Kathryn Paulk

Other Books by Kathryn Paulk

- Algebra 1 Help
- Algebra 2 Help
- Pre-Calculus and Trig Help
- College Algebra Help

- Calculus 1 Review in Bite-Size Pieces
- Calculus 2 Review in Bite-Size Pieces
- Calculus 3 Review in Bite-Size Pieces
- Differential Equations With Applications: Class Notes With Examples

---

- One-Page Summaries for Algebra, Geometry, and Pre-Calculus
- Graphing Functions Using Transformations for Algebra & Pre-Calc.
- Complex Numbers and Polar Curves For Pre-Calc and Trig: With Problems and Detailed Solutions
- Discrete and Continuous Probability Distributions: A Creative Comparison (V2)

- Teach Your Child to SWIM

BIG MATH For Little Kids

Workbooks for Young Children
& Solution Manuals for Parents

- Introduction to Numbers

- Introduction to Fractions
  by Sharing Things

- Introduction to Counting and Fractions
  by Cooking Breakfast

- Learn About Fractions   *****
  by Baking Cookies

- Adding Big Numbers, Guessing Numbers
  and Secret Codes

- Learn to Graph by Riding Bikes
  on Graph Paper

Made in the USA
Monee, IL
08 January 2025

76388567R00075